Anonymous

Observations on Many of the Active Medicinal Substances of the Vegetable Kingdom

especially in relation to their collection, preservation, and preparation for

use

Anonymous

Observations on Many of the Active Medicinal Substances of the Vegetable Kingdom
especially in relation to their collection, preservation, and preparation for use

ISBN/EAN: 9783337184087

Printed in Europe, USA, Canada, Australia, Japan

Cover: Foto ©berggeist007 / pixelio.de

More available books at **www.hansebooks.com**

ON MANY OF THE ACTIVE

MEDICINAL SUBSTANCES

OF THE

VEGETABLE KINGDOM;

ESPECIALLY IN RELATION TO THEIR

COLLECTION, PRESERVATION, AND PREPARATION
FOR USE.

Not Published.

BOSTON:

PRINTED BY JOHN WILSON AND SON,

5, WATER STREET.

1864.

MEDICINAL SUBSTANCES

OF THE

VEGETABLE KINGDOM.

THE country practitioner of medicine should qualify himself to act in the double capacity of physician and apothecary. It is not necessary that he should study and make himself fully acquainted with all the duties of the dispensary, or to become master of many shop-manipulations, which are, at this day, considered essential to the city dispenser of medicines. He need not learn to gild his pills ; to fold his powder-potions in chartulas of an exact length, breadth, and thickness ; to enclose each acrid, balsamic dose in capsules of gelatine ; or to compound the popular perfumes and skin-beautifying lotions, which, at the present time, constitute an important item in the sales and profits of the city apothecary. But the country physician should learn all the great truths of the *Materia Medica*, and understand the leading principles of pharmacy, as well as the thoroughly educated apothecary.

The physician's knowledge of chemistry, which has become an important branch of medical education, will enable him to determine the purity of those medicines which result from chemical combinations ; but this branch of his professional training will afford him little assistance in ascertaining the nature of those simple medicinal remedies which are derived from the vegetable kingdom. He should be able to decide, when he purchases his barks, roots, seeds, flowers, and fruits, his *cinchona, ipecacuanha, colchicum, arnica,* and *colocynth,* whether they have been well cured, and are in their full virtue. He should, moreover, learn how to preserve them, as far as possible, in their entire activity; and before they begin to decline in their efficiency, from the changes of temperature, the humidity of the atmosphere, the contact of air, the influence of light, or from any circumstances which are incident to age, he should understand how he may secure their full powers by a form of preparation which will extract and retain their entire medicinal qualities for an indefinite length of time, if such form of preparation exists, and especially if it may be considered an eligible form for extemporaneous use.

But the country physician, who studies two professions, at an increased expense of time and money, will receive, in after-time, a substantial compensation for the additional labors which were required of him in his preparatory course of study. If he selects his own simples with good judgment, and compounds them in a skilful manner, he will understand their effective powers ; and, should his prescription fail of producing the expected result, he feels certain wherein he has made a mistake. He knows it is not on account of any diminished power in his recipe, but that it results from the imperfect estimation he had formed of his patient's disease, or of his constitutional susceptibilities. He studies his case again with the devotion of all his powers; and, after making a new *diagnosis* of the disease, he alters his prescription to meet its wants.

The city physician is relieved from the necessity of studying a twofold profession. He usually relies on the ability and fidelity of his apothecary; and, in the large cities, he may ever be able to select those in whom he has full confidence. But embarrassments sometimes attend city practice, which the country physician, on account of his peculiar course of preparation, does not experience. The city practitioner occasionally orders a recipe, which is compounded by the apothecary, and which is taken by the patient in the prescribed quantity and at specified intervals, but which fails to produce any visible effect. When cases of this kind happen, he is necessarily in doubt whether he misapprehended the requirements of his patient, or whether his prescription failed on account of the impaired condition of some of its constituent parts, or from a possible error in its composition. His mind is turned alternately in two directions; viz., to the quality of the recipe he has ordered, and to the character of the disease, together with the constitutional requirements of his patient. He hesitates between two contingencies; while the country physician, who is his own apothecary, is relieved from this dilemma of his professional brother, and left entirely free to consider his patient's case.

Most medicines are administered in combination, or in compounds; and this is more especially true of those which are usually employed in chronic diseases. By compounds, in these cases, we do not mean those which result from the union, in a given degree, of an acid with a metal, with an alkaline earth, or a vegetable substance, but an admixture of simple remedies in the form of *powders, pills, infusions, syrups, decoctions, tinctures,* &c. The metallic compounds, the neutral salts, the sulphates, muriates, and carbonates of vegetable basis, are principally prescribed in acute diseases; and the simple or vegetable remedies sometimes *per se,* but more generally in composition, in complaints of a chronic nature. This seems to be the rule of practice, although there are many exceptions to its operation, and perhaps too many for the confirmation of the rule.

The simple or natural medicines, described and recommended in our present systems of *Materia Medica,* by far exceed in number the artificial remedies, or those which result from the operations of chemistry; and it is much more difficult to obtain these vegetable, medicinal substances of the highest quality, and to preserve them in their full efficiency, than it is to procure and preserve the best chemical compounds. Complaints are made, and no doubt with reason, of the adulterations of *mercurials, antimonials,* and other preparations, more especially the metallic; but the substances added to increase the weight of these are usually in themselves inoperative: yet, as there seems to be no rule in the extent of these admixtures or adulterations, physicians are frequently embarrassed by a knowledge of the uncertain strength of these articles. But such as these remedies are, when they come from the laboratory of the manufacturer, such they remain, with comparatively slight care, for a long period. The physician who orders a new supply of *calomel* or *emetic tartar*, and puts the same into well-stopped bottles, and properly secured from the effects of light, after having once ascertained their strength, may thereafter prescribe of the same with confidence as to their operative powers, even to the end of his supply. The writer once had occasion to give an antimonial emetic from the few grains that remained in the phial of a physician who had retired from practice thirty years previous to that period; and the operation which followed, was, in all respects, effectual and characteristic.

But there is no such permanence in the strength of our vegetable medicines. When obtained and cured in a proper manner, some decay of their virtue at once becomes incident to them, even under the greatest care; and, without such care, the early loss of their active properties is nearly certain.* Nor is this by any means the whole difficulty in the case. The plant

---

* "Some remedies, and generally those of a vegetable nature, lose much of their activity by age, and therefore require to be given in large doses to produce the desired effect. Hence, where a physician has been induced to increase the usual dose of some article, which in a recent or unaltered state is active, and even poisonous, but from age, or other deteriorating causes, has become weakened, serious accidents may occur if he persist in administering the same quantity of a parcel which is fresh, or which is obtained from another apothecary." — *Thomas's Universal Formulary,* p. 62.

from which the bark is obtained, or from which the
root is taken, should neither be too old nor too young.
In other words, the several parts should be taken when
such parts are in their full vigor. The season of the
year is to be consulted in many cases. The roots should
never be gathered while the stalk is in full growth;
for, in such cases, the life, or the peculiar quality of
the radical portion of the plant, ascends, in a modified
state, to the stalk, leaves, flowers, and fruits, or seeds.
The quality of the soil in which the plant grows should
also be considered. Thus *helleborc* and *skunk-cabbage*
should be taken from low and wet ground, which seems
to be their native and proper soil. When these grow
in the higher and warmer soils which border the damp
meadows, they develop a larger stalk and leaf; but the
roots lose, in a great degree, their peculiar fetid and
medicinal character. On the other hand, *snakeroot*
in its several varieties, *valerian*, *gentian*, &c., attain
their highest medicinal perfection in a higher, warmer,
and richer soil; and if perchance they should be taken
from land of an opposite character, where they sometimes grow, the specific odor and quality of the roots
would be found to be comparatively weak and inefficient.*

---

* " Some herbs, in their infancy, abound most with odoriferous matter, of
which others yield little or none until they have attained a more advanced
age. . . . The roots of some of our indigenous plants, whose juice is,
during the summer, thin and watery, if wounded early in the spring, yield
rich balsamic juices, which, exposed to a gentle warmth, soon concrete
into solid gum-resins. . . . In open exposures, dry soil, and fair, warm
seasons, aromatic plants prove stronger and more fragrant, and fetid ones
weaker in smell, than in the opposite cases. To these particulars, there-

But it is not enough to collect from the plant in its proper age, from a congenial soil, and at the right season. The articles thus gathered should be cured or dried in such a manner as to preserve, in the highest possible degree, their full medicinal virtues. They should be dried sufficiently to prevent subsequent mould, when properly packed in casks or boxes, but not so far beyond that degree as to render them friable, inodorous, and thus comparatively valueless. The drying should be effected in a temperature not over 80°, and in the shade. If slippery-elm bark — sometimes employed in medicine on account of its mucilaginous quality, and frequently as a luxury, on account of its

---

force, due regard ought to be had in the collecting of plants for medicinal use. . . . It may be proper also to observe, that the different parts are often very different in quality one from another. Thus the bitter herb *wormwood* rises from an aromatic root, and the narcotic *poppy-head* includes seeds which have no narcotic powers." — *New (London) Dispensatory*, p. 4.

" Let the annual roots be gathered before they shoot out their stems and flowers, — the biennial principally in the autumn of that year in which the seeds are first sown; and the perennial when the leaves begin to fall, and therefore, generally, in the autumn. Having first washed away the filth, and cleared them of their withered and corrupt fibres, hang them up in a *shady, airy place,** that they may dry moderately. Let the thicker be cut in pieces, either lengthwise or transversely; preserving the cortical part, and rejecting the pith. Those roots which lose their virtue by drying should be kept covered with dry sand. Let herbs be gathered at the time of their vigor, when they shoot into perfect leaves, but before the flowers are opened. Of some, it is best to take only their flowering tops. Let them be dried as directed above with respect to roots. Let flowers be gathered when they are moderately expanded, upon a clear day, before noon. Let seeds be gathered when ripe, and beginning to dry, before they fall spontaneously. Woods for medicinal uses are best felled in the winter; and this is the best season for shaving off the bark."—*Pharmacopœia Universalis, Lond.; second edition*, p. 448.

* The Italics, generally, in our extracts from various authors, are our own.

pleasant taste, fragrant odor, and crisp substance — be taken from the tree in the warmest season of the year, and exposed to the sun's rays, it dries quickly, turns of a dark color, loses much of its pleasant odor, becomes tough and stringy, and is of little market value.

Valerian root, taken from the earth and dried in the sun, loses a large share of its essential and valuable volatile qualities, and soon becomes friable, tasteless, and worthless.*

But we need not speak of the proper method of curing the simples of the *Materia Medica* singly, because a general rule applies to them all when under collection for medicinal purposes : viz., that they should be dried in the shade, in a moderate atmospheric temperature ; so .far dried as to prevent the danger of rust or of decomposition, and still not over-dried so as to deprive them of their active properties. This rule is important in all cases, but more especially so with those articles whose principal virtue resides in their volatile oil and resinous juices. When these substances are properly dried, they should be packed in such a manner as to protect them as far as possible from the contact of air, and from the influences of an atmosphere alternately damp and dry.†

---

* "The root [valerian], if not taken up at a proper time and properly preserved, becomes inert." — *Parr's Medical Dictionary.*

† "The proper preservation of medicines is an object of the greatest importance to the apothecary. The aromatic gums and resins should be kept secluded from the light, and, as much as possible, from the air, in perfectly dry rooms. Boxes and barrels, with tight covers, will serve for holding barks and roots after they have been thoroughly dried. Roots and bulbs, such as liquorice and squills, which are to be preserved fresh, should be buried in dry sand. Leaves and flowers should be kept in dry canis-

But the greatest care that can be bestowed in the preparation of vegetable medicines, and for their preservation to the period of their use, is too often rendered unavailing by subsequent contingencies in passing through several hands before their prescription for immediate use. Even if these articles are properly cured and packed, they are, of necessity, frequently opened by the druggist, and sold in small quantities to the apothecary and physician ; and the bales, casks, and boxes thus opened cannot in all cases be at once closed and properly secured. And, with all possible care in this respect, these frequent exposures to atmospherical influences cause the work of deterioration to commence. The apothecary and physician keep their supplies thus purchased in drawers, or perhaps in paper parcels, which are frequently necessarily exposed to the air and light ; and the inevitable consequence is, that the contents of their shops, thus kept, must be gradually declining in value and efficiency. We do not here speak of powders, which may be better preserved in well-secured "specie-bottles," or of liquids kept in "glass stoppers."

Let us illustrate our meaning by a few examples. *Chamomile flowers* come to us in large casks, closely

---

ters. . . . The apothecary should frequently examine the condition of every article; and, on the slightest appearance of mouldiness or the attack of insects, should clean them, and again dry them *in a heat from seventy to a hundred degrees.* This examination and redrying, which should be made several times in the year in respect to the articles which are most likely to change, should especially be made early in the spring, of all roots, barks, and leaves in the shop; and those of which the sensible properties have become impaired should be rejected." — *United-States Dispensatory,* p. 753.

pressed, and secured from the action of air. When they are first opened, they are entire in their several parts, bright, highly aromatic to the taste, and of strong though grateful odor. A part of these flowers are soon removed from these large packages, weighed, and put in small casks and bundles, from which they are transferred to the drawers of the apothecary or physician. They are still very handsome and strong-scented, still nearly perfect in their form : but after a few months, or even weeks, the color becomes dull, the flavor less obvious, the form of the flower less distinct ; the process of decay still going on, until they partially fall into dull gray powder, and lose all their medicinal activity, except, perhaps, a part of their astringency.

Sarsaparilla, which has been well cured and packed, comes to our market in flexible roots, tough, and even withy. If the root be split, it appears bright and entirely sound ; and, if a small quantity be chewed, it leaves in the mouth a peculiar acrid taste. If the roots be sent to market before they are properly dried, they are less flexible ; and, when opened, the interior looks dull, and perhaps slightly mouldy, and the taste is less acrid. But the good root, perfectly dried and however carefully kept, gradually deteriorates, and at last becomes brittle, tasteless, and worthless ; still appearing externally very much as it did in its best condition.*

In view of these facts, can we wonder why different opinions have been so long entertained of its virtues by

---

* " The sarsaparilla of the shops is very apt to be nearly or quite inert, either from age, or from having been obtained from an inferior species of smilax." — *U. S. Dispensatory,* p. 638.

the members of the medical profession? Many physicians esteem it of great value, and speak with the strongest confidence of its efficacy in some of the most deep-seated and obstinate diseases for which they are called to prescribe. The " UNITED-STATES DISPENSATORY " devotes six pages to the history of this plant, and the description of its medicinal virtues; and, in certain past periods, it has been in great favor and in extensive use. At other periods, medical men have almost unanimously pronounced it worthless. The " NEW AMERICAN DISPENSATORY " disposes of sarsaparilla in nineteen lines; declaring it "a very inert, mucilaginous substance." Dr. CULLEN condemns it almost in a single line; pronouncing it " wholly inert." In PEREIRA'S " MATERIA MEDICA," seventeen royal octavo pages are given to its history and the proof of its virtues. Thus, for two centuries, it has fluctuated among medical men, from the highest favor and most general use to the most manifest disfavor and neglect.

How can this conflict of opinion in relation to the worth of sarsaparilla as a remedial agent, which has existed ever since the days of the celebrated VESALIUS, be explained? Easily, and in this wise : Those who have used it with success have taken the root in its best form, have prepared it in a way to secure its full strength, and administered it with proper adjuvants. Others, who consider it unworthy a place in the *Materia Medica*, have had the misfortune to employ the roots whose virtue had departed; or perhaps have taken them in their virtue, and, by their mode of preparing

2

them for use, have unconsciously dissipated their active,
most valuable principles. A similar conflict of senti-
ment would exist among medical men, if a part of the
profession were to give opium, pure and nearly as inspissated about the incision of the poppy-head, and another
portion should use the extract made from the capsules;
or if some were to form their grain-pills from the
mass of the best commercial opium, while others
should insist on taking the same kind of opium, and
preparing it for use by torrefying it on a hot shovel!

Perhaps there is no article of the *Materia Medica*
which requires more care in its gathering, cure, and
preservation, than *colchicum root*. When administered
in its full powers, there can be no doubt of its being an
active, remedial agent. The bulb is often taken and
used in the fresh state in the countries where it is produced; and, when taken from the earth *at the proper
time*, its taste is bitter, hot, and acrid. But it is a very
difficult matter to dry the root, without the escape of its
volatile and active properties. Sometimes the bulb is
partially dried in the shade, in the temperature of the
atmosphere, and packed in dry sand, in which it is
imported into this country. In this way, its efficacy
is often preserved for a considerable time. At other
times, the root is imported in a thoroughly dried
state, appearing bright and sound, but which is nearly
inodorous, tasteless, and inert.* The testimony in rela-

---

* " The medicinal virtue of the bulb [*colchicum*] depends much on the
season in which it is collected. Early in the spring, it is too young to have
fully developed its peculiar properties ; and, late in the fall, it has become

tion to the worth of this medicine is as conflicting as in the case last named. One person (Kraff) declares that he ate whole bulbs, without experiencing any inconvenience; and it is stated, that, in some countries, the peasants use it for food when collected in autumn. On the other hand, it is declared that there is abundant testimony of its highly irritating and active medicinal character. Like sarsaparilla, it is at one time a most popular remedy; and, at another period, its reputation subsides into disuse and general disesteem. It was used by the ancients as a *diuretic*, and in cases of gout. STÖRCK revived its use in more modern times. He recommended it as a diuretic in dropsy, and an expectorant in asthma; and, for another period, it had a high reputation for these complaints. But its uncertain operation again led to its general abandonment and seemingly final disuse. It was again brought into notice by a *London* physician, Dr. WANT, who attempted to prove that it was the active ingredient in the *eau médicinale* d'HUSSON, — a very celebrated preparation for the cure of the gout. It has, during the present century, been fluctuating in favor. By some physicians, it has been highly commended for the cure of gout, rheumatism, neuralgia, &c.; and by others it has been pronounced utterly inert. When colchicum

exhausted by the nourishment it has afforded to the new plant. The proper period for its collection is from the early part of June, when it has usually attained perfection, to the middle of August, when the offset appears. It is probably owing, in a great measure, to this inequality in the *colchicum* at different seasons, that entirely opposite opinions have been given by different authors of its powers." — *U. S. Dispensatory*, p. 256.

root, in its full powers, has been judiciously administered, it has, undoubtedly, proved a most efficacious and valuable remedy in many complaints, and on this account has, in repeated instances, risen to great favor ; but the frequent use of the inert root, and the consequent disappointment of the physician in its employment, has thus far prevented its lasting popularity with the profession.

If the preservation of vegetable medicines in their simple forms be a matter of so great difficulty, it becomes very important to learn how they can be taken in their best estate, and be compounded or prepared in such a manner as will permanently secure their full activity ; due regard being given in such preparations to the forms in which they should be administered to the patient.*

Some prescriptions will disturb the patient by their unpleasant taste or odor, while other forms of the same remedy will prove in no wise offensive, and may conduce to his immediate comfort and composure. Thus, powdered *aloes*, or *jalup*, given in any of the usual extemporaneous vehicles, may create a disturbance and disgust in the system, in some cases not easily removed ; when the same substances, in the pill form, might cause no unpleasant sensation. If an infusion or decoction of

---

* " The form in which medicines are exhibited is often an object of considerable importance. By variation in this respect, according to the nature of the medicine, the taste of the patient, or the condition of the stomach, we are frequently enabled to secure the favorable operation of remedies, which, without such attention, might prove useless or injurious." — *United-States Dispensatory*, p. 1318.

Canada snakeroot be taken, even on its first preparation, but more especially after the *menstruum* has lost the aroma of the root and becomes vapid, as it frequently does in a few hours, it will, in nearly every case, prove more or less offensive to the taste, and may provoke nausea and vomiting. If, instead of the infusion, we should give a teaspoonful of the tincture, prepared from the fresh root, it would almost invariably prove pleasant to the taste, tend to relieve nausea (if any existed at the time), and produce, to some extent, its characteristic, cordial, and stomachic effect.

But forms of prescription should not be made agreeable to the taste, at the expense of the essential qualities of their component parts. The great desideratum with the physician is to ascertain the best manner of composing his *recipes* so as to secure the entire virtues of their constituent parts, and in such a manner as will permanently preserve their full efficiency for " an indefinite time ; " and thus to have in reserve compounds not subject to decline in force, instead of simple vegetable remedies, so liable, even under the greatest care, to early decay. This consideration being paramount with him, he may study eligibility of forms as far as the more important consideration of efficacy may admit.

We will now consider briefly some of the authorized preparations of vegetable medicines for immediate use. The most common *formulæ* are infusions, decoctions, syrups, inspissated juices, fluid extracts, powders, pills, and tinctures.

" *Infusion* is a term employed in *pharmacy* to denote

2*

the operation in which water, on remaining for some time on vegetable matter, dissolves part of it; and also to express the preparation which results from that operation." — *New American Dispensatory.* Thus we have infusion of bark, foxglove, gentian, rhubarb, snakeroot, valerian, senega, &c. These infusions are sometimes made with cold water, and sometimes by the use of boiling water. Those made with cold water, with a very few exceptions, extract only to a slight extent the active principles of the substances employed. Boiling water dissolves more of the *albumen, mucilage,* and *starch,* but leaves almost untouched the essential medicinal qualities of articles submitted to its influence. *Infusions* very soon *change,* and become offensive. Sometimes, in warm weather, they are rendered vapid and nauseous in a very few hours after they are prepared. In their best condition, they are of little use; and so general is the acknowledgment by the profession of this fact, that it appears somewhat surprising that their numerous forms are still preserved in our modern dispensatories.*

Decoctions differ from *infusions* simply, as the former

---

* " The *infusion* [of medicines in water] is, however, liable to one very great objection, — that it cannot be kept, *even a very short time, without being decomposed or spoiled.*" — *American New Dispensatory,* p. 186.

"As infusions do not keep well, especially in warm weather, they should be made extemporaneously and in small quantities. . . . *The propriety of their introduction into the* PHARMACOPŒIA *has been doubted.*" — *United-States Dispensatory,* p. 1007.

" Infusions . . . are exceedingly liable to decomposition; and, consequently, cannot be kept long without spoiling." — *Thomas's Universal Formulary,* p. 520.

term implies. Vegetable substances are *boiled* in water for a longer or shorter period, with a view of extracting their active principles for medicinal use. These *decoctions* are generally employed to form the bases or active properties of *syrups*; and the continued boiling, sometimes ending in evaporation, or "simmering down," to produce the *extracts*, solid and fluid. In all our PHARMACOPŒIAS, there are many directions given for *decoctions*, notwithstanding the frequent confessions running through their pages, that they very imperfectly extract and retain the most valuable medicinal agencies contained in vegetables. The remedial virtues of the *vegetable kingdom* reside almost wholly in the *aroma, essential oil, resin* and *tannin*, contained in the several valuable varieties of that department of medicine. Boiling water, in time, disengages these principles, and dissipates them in the rising vapor. None of them are permanently retained by the water, except the *tannin*, or astringent principle; and this, as it will appear, generally in a state of at least partial decomposition. The extracts which are retained by decoction are mucilage, starch, albumen, salts, &c.; none of which are esteemed as strictly medicinal.

It is difficult to understand why these numerous forms of decoctions are still preserved in the books, which so freely acknowledge the utter failure of such prescriptions to extract and preserve the virtues of the medicinal substances enumerated in our systems of *Materia Medica*. Still, increasing numbers of this form of preparation are given in our modern dispensatories, and some of them printed on the same page in which their

eligibility and propriety are denied.* The dose of decoctions varies greatly, as prescribed at the conclusion
of each formula, and as stated by different authorities.
The dose sometimes ordered, if the preparation dis-

---

* The effect of boiling differs from that of infusion in some material
particulars. One of the most obvious differences is, that as the essential
oils of vegetables, in which their specific odor resides, are volatile in the
heat of boiling water, they exhale along with the watery steam, *and thus
are lost to the remaining decoction.*" — *New* [*London*] *Dispensatory*, p. 278.

"In the heat of boiling water, the essential oil of vegetables, *in which
their virtue generally resides,* is dissipated. . . . The grosser parts of many
substances are only extracted by boiling. . . . Boiling water extracts, for
instance, the rougher and disagreeable parts of chamomile and the *carduus
benedictus.*" — *Parr's Med. Dictionary,* art. " COCTIO."

. . . "And as plants lose by boiling all that which goes off by heat, in
the form of vapor, with 212° of heat, all those plants are unfit for this operation whose virtue required is volatile oil. . . . Those whose virtues reside
in a more fixed matter (astringency) are fit for decoction. . . . Let it, however, be carefully observed, that the peculiar virtue of a plant, which can
only reside in its presiding spirit, does not always show itself by some
remarkable odor, fragrance, or aromatic taste. On the contrary, it may
happen that the spirit may be extremely active, without remarkably affect-
ing the senses." — *Pharmacopœia Universalis* (*Lond.*), p. 510.

"By boiling vegetable substances in water, the active matter is more
abundantly dissolved than by infusion. The preparation thus obtained is
called a decoction. . . . Though a large portion of matter is dissolved by
the water by this mode of preparation, yet it cannot always be advantage-
ously employed. Whenever the virtue of the substance subjected to it
depends, *in whole or in part,* on any volatile principles, they are necessarily
injured by being thus dissipated. . . . Many vegetables suffer injury by
boiling, even when this cannot be ascribed to the dissipation of their volatile parts." — *American New Dispensatory,* pp. 495, 496.

" Decoction, or boiling, is sometimes employed in extracting the virtues
of plants; but it is often disadvantageous, *as most of the proximate principles of vegetables are altered by it,* especially if long continued." — *United-
States Dispensatory,* p. 762.

" All vegetable substances are not proper objects for *decoction.* In many,
the active principle is volatile at a boiling heat; in others, it undergoes
some change unfavorable to its activity; and, in a third set, is associated
with inefficient or nauseous principles. . . . *Decoctions,* from the mutual
re-actions of their constituents, as well as from the influence of the air, are
apt to spoil in a short time." — *United-States Dispensatory,* pp. 899–900.

solved and retained the full virtue of the ingredients, would seem to be objectionable ; nay, almost incompatible with the continued vitality of the patient : but, weakened as they are by this mode of preparation, the

---

" By ebullition in water, the volatile constituents of vegetables are dissipated; and hence, when these are the active principles, *the process is an objectionable one.* It is obvious that the *saffron*, in the *decoctum aloës compositum*; the *sassafras*, in the *decoctum sarzæ compositum*; and the *juniper*-fruit, in the *decoctum scoparii compos.*, — are deprived of their volatile oils by boiling; and therefore these preparations are, on that account, objectionable." — *Pereira's Elements of Mat. Med.*, vol. i. p. 305.

" An objection has been taken to this [*decoction of sarsaparilla*], as well as to all other preparations of sarsaparilla, by boiling water. An *infusion of sarsaparilla*, says SOUBEIRAN, which is odorous and sapid, *loses its odor and taste* by boiling *for a few minutes.* *These changes speak little in favor of the decoction.*" — *Nouv. Traité de Pharm.*, vol. ii. p. 168.

" With respect to *bitters* [by ' bitters ' are meant the tonic principles in vegetables; see vol. ii., p. 39, of the authority cited below], it is certain that *decoction* extracts more powerfully than infusion: *but by dissipating any aromatic parts that are joined to the bitter*, and by extracting more of the earthy part, and what may be called a coarser *bitter*, *decoctions* are almost always more disagreeable than infusions; and therefore what we call *extracts*, which are always prepared by decoction, are always less agreeable to the stomach than the bitter substance. It appears to me that *decoction decomposes the substance of what is extracted*; for it is seldom that *decoctions* do not, upon cooling, deposit a part of what they had suspended before, and that, also, a matter different from the entire substance. . . . It is pretty certain that *bitters* are never treated by decoction so as to be either agreeable or very useful." — *Barton's* [*Cullen's*] *Mat. Med.*, vol. ii. p. 49.

. . . " Many substances are injured or destroyed by *decoction*, especially when their active principles are volatile, and when, during ebullition, chemical changes take place, by which the active constituents are rendered insoluble or decomposed." — *Thomas's Universal Formulary*, p. 520.

" *Decoction* is sometimes employed in extracting the virtue of plants; but it is often disadvantageous, *as most of the proximate principles are altered by it.*" — *United-States Dispensatory*, p. 762.

. . . " Mais cette forme [*décoction*] a l'inconvénient d'être ordinairement désagréable à la vue, et au gout: au reste on ne s'en sert point dans le cas urgens, parcequ'elle ne peut pas s'exécuter avec promptitude." — *Encyclopédie des Sciences, par* M. DIDEROT.

greatest danger would seem to be from the quantity.
Four pints a day of the decoction of sarsaparilla, might,
as we think, offend a weak digestion. We subjoin the
recipe in which this bulky dose is prescribed : —

  " **R**. *Sarsaparilla*-root, cut in bits, two ounces. Macerate for a day in six pints of water ; after which, boil over a
gentle fire, in a double vessel, with a closed lid, . . . until
one-third or one-half be evaporated. Of this decoction, the
patient is to take very early, in bed, a glass that will hold
ten ounces : what remains serves for the rest of the day.
This course is to be continued for twenty days." — *Pharmacop.
Universal (Lond.)*, p. 281.

In the above recipe, the directions are to boil " in a
double vessel, with a closed lid," until the liquid should
be reduced to one-third or one-half, or until the six pints
should be evaporated to three or four pints. However
the vessel may have been *doubled*, and the *lid closed*, one-
third at least of the water must be dissipated in vapor ;
and, with this vapor, all the essential virtues of the sarsaparilla will escape, as we think will be shown in the se-
quel. Boiling "in a close vessel," or in any other manner
in which the quantity of the menstruum is to be reduced
by evaporation, amounts to the same thing as boiling in
an open vessel. All the different and somewhat inge-
nious appliances which have been made to obtain the
most valuable solutions by the agency of boiling have
failed of answering any valuable purpose. " Papin's
Digester" * was an early contrivance to effect this ob-

---

* " An instrument invented by Mr. PAPIN. It consists of a strong vessel of copper or iron, with a cover adapted to screw on with pieces of felt
or pasteboard interposed. A valve, with a small aperture, is made in the

ject; but this mechanism has long been in entire disuse. Boiling *in vacuo* is spoken of, at the present day, with some favor. Let us see, for a moment, how this would operate in practice. If a strong, air-tight, metallic vessel were half filled with water and medicinal substances, and the air should be withdrawn from the upper half, that part would immediately fill with vapor, upon the application of heat, which should raise the temperature of the liquid to the boiling-point, and there would no longer be a *vacuum.* Now, if this vapor be withdrawn by the constant use of the pump, as has been proposed, the volatilized essential virtues of the medicinal plants employed in the case would constitute a part of the fluid thus removed. It might as well have risen in the vapor from an open vessel. If the vapor be not withdrawn during the process, the essential and valuable qualities of the plants would be decomposed or destroyed. That such would be the case, we think we have shown by the authorities already cited, even when the decoction has been made in an open vessel, in a temperature elevated only to 212° Fahrenheit. How much more certain and complete would be the decomposition with a temperature of 400°! Admitting, however, the possibility that any component essential part might remain in the decoction, in an unaltered state, when withdrawn from the vessel, — a mere possi-

---

cover; the stopper of which valve may be more or less loaded, either by actual weights, or by pressure from an apparatus, on the principle of the steelyard. The purpose of this instrument is to prevent the loss of heat by evaporation. Water may be thus heated to 400° Fahrenheit." — *Encyclopædia Americana.*

bility indeed, — it would not mingle and remain in the cold menstruum.   The volatile oil, aroma and resin, the most valuable qualities of vegetable medicines, *will not unite with water.*

Heat is the great thief of our vegetable *Materia Medica.* Its external application seems to be fatal to the internal life or active force of all those plants which are most esteemed as remedial agents. This, we believe, may be considered as the rule. There may be some seeming, possibly real, exceptions to its operation ; but they will not be found more than sufficient to establish its correctness. How explicit and imperative are our authorities to avoid heat in the cure and preservation of plants ! " Let them be carefully dried *in the shade*, in a temperature of 70°, or above ; but *in no case above* 100°." " When cured, let them be preserved in a *cool*, dry place ; " &c. Still, in the same authors, how frequent are the directions to subject the roots, barks, seeds, and flowers, thus collected and preserved, to a temperature of 212°, which some have recommended should be increased even to 400° ! With few exceptions, all the book-rules for the preparation of decoctions, syrups, and extracts, — and the prescribed forms of these compounds are quite numerous, — require the boiling of their several constituents in water ; which boiling is to be continued until the water is reduced in measure from one-third to one-half of the quantity first taken.

Heat is truly the thief of the volatile and essential qualities of vegetable medicines, notwithstanding the seeming contradiction presented in a very few instances. *Coffee*, which may be claimed as a medicine, is prepared for use by *roasting*. But it is by no means certain that coffee loses any of its essential strength by this process. It does lose, by this heating, certain qualities not desirable for table use. Old coffee loses in weight, and old coffee requires less roasting. The best solution which chemists have been able to give of the effects of this roasting is substantially as follows : That the aromatic, volatile oils of raw coffee are in minute particles, in the small cells of the berry, or seed, *enveloped in a fatty substance, almost or quite impermeable;* that the roasting dissolves and eliminates this tenacious fat, and releases these volatile principles, which, being thus released, soon pervade the whole substance of the berry, and cause it to increase in size ; which enlargement of form is the evidence that the roasting process is completed. This is the point of *" bay-browning;"* and, if the roasting be continued beyond this point, the flavor and strength are weakened, and sometimes destroyed. Thus it is evident that coffee should be withdrawn from its *" browning "* as soon as its active principles are released from their cellular confinement and diffused through the substance of the berry, *and before any of its valued properties are vaporized and lost.*[*]

---

[*] " The aromatic, volatile oils of raw *coffee* are tenaciously retained by the fatty oil. They undergo alteration of property by the operation of roasting. . . . The chemical changes effected in *coffee* by roasting require further investigation. . . . If coffee be over-roasted, either by employing too high a temperature or by carrying on the process too long, its flavor is greatly impaired."— *Pereira's Elements of Mat. Med.,* vol. ii. p. 630.

" The most important principles [of coffee] are the *caffeic acid,* resembling, in its astringent character, the tannin of *tea;* . . . and the fragrant, volatile oil, called *caffeone.* This oil is distinguished by the microscope in minute drops, in the cells, or between the outer membrane and the body

The heat of distillation vaporizes the essential oils of
the mints and of other vegetable substances; but the
oils thus volatilized no sooner rise from the heat of
212° than they pass a condensing temperature, redu-
cing the heat nearly or quite 150°, by which they are
brought into their officinal, liquid form. This rising
vapor does not touch the atmosphere for an instant
— except the small portions of it which are allowed to
escape to prevent explosion — until after it takes the
form of oil, and is removed from the "*receiver.*" If
such an exposure should be allowed, even for a moment
of time, there is no known process — no seeming possible invention — by which the vapor could be re-
collected and reduced to the form of oil, or be returned
to the menstruum in the "*retort.*" If once exposed to
the atmosphere, the volatile principle is instantly diffused and lost.

These essential oils, of the mints especially, are not
necessary to the physician. He obtains, although
not in so concentrated a form, all the virtues of these
simple vegetable substances by a tincture made from
the fresh or well-cured plants. They are far more
important to the confectioner than to the apothecary.

Nor are these oils secure from changes induced

---

of the seed. Roasting *disperses this through the solid substance,* and in part,
or wholly, expels it, if carried too far." — *New American Cyclopædia.*

" *Coffee* improves by age, losing a part of its strength, and thus acquiring
a more agreeable flavor." . . . It contains (among other things) " thirteen
per cent of fatty matter, and fifteen of *glucose.* . . . It loses [by roasting]
about twenty per cent of its weight: it acquires at the same time a peculiar
odor, entirely different from the unaltered grains, and a decidedly bitter
taste. . . . If the roasting be too long continued, it renders the coffee
unpleasantly bitter and acrid." — *United-States Dispensatory,* p. 1250.

by a moderate heat, and exposure to the atmosphere. Heat and air are still their enemy. The confectioner or the apothecary sometimes opens a canister of the oil of peppermint, which was imperfectly corked, and which, perhaps, had stood on a shelf, next a brick wall of southern exposure, or near a window admitting the sun; and finds, instead of a delicately flavored essence which might have cost him some four or five dollars the pound, a mere terebinthinate oil, worth scarcely as many cents, apparently differing in no respects from the turpentine distillations of North Carolina. These losses seldom occur twice to the same individual. They teach a lesson too expensive to be easily forgotten; and he who has suffered once in this respect is ever thereafter careful to place his essential oils, especially the *mint* oils, well secured from the air, and in the coolest place his premises afford.

Water-baths have been recommended as possessing some advantages over decoctions made by an open fire. "The water-bath is to be used in all cases in which a heat above boiling water would be injurious. When a temperature above 212°, and not exceeding 228°, is required, the water-bath may be filled with a saturated solution of common salt."

Some have supposed and asserted, that the water-bath possessed decided advantages over the ordinary modes of boiling; but pharmacologists do not generally claim this advantage. The fact is, the vaporization of water, however caused, is substantially decoction. If "eight pints of liquid are to be gently evaporated to twenty ounces," one hundred and eight ounces of the one hundred and twenty-eight are to be dissipated by heat in the form of vapor; and thus all the unfavorable effects of more rapid boiling are equally attendant on this slower process of evaporation.*   In fact, no substantial advantages are now claimed for the water-bath, except that, "*in simmering down*" the decoction to a solid extract, there is less danger of burning it.

The operation of heat on the curative properties of vegetable substances is clearly illustrated from its effects on *nux vomica*.   The action of this medicine, when prepared in its full strength, is specific and uniform.   In Pereira's "Materia Medica," vol. ii. p. 538, we read as follows : " On the vertebrata its effects are very uniform, though it requires larger quantities to kill herbivorous than carnivorous animals."   But we find, in another current authority, that its operation "is very uncertain."   The " United-States Dispensatory " says, page 479, "*Nux vomica* may be given in powder, in the dose of five grains, repeated three or four times a day, and gradually increased till its effects are experienced.   In this form, however, it is very uncertain ; and fifty grains have been given with little or no effect ! It is most readily reduced to powder by filing or grating ; and the raspings may be rendered finer by first steaming them, then drying them by a stove-heat, and,

---

* "Evaporation by means of the water-bath, from the commencement of the process, is safer than the plan just mentioned [rapid boiling], as it obviates all danger of burning the extract: but, as the heat is not applied directly from the fire, the volatilization of the water cannot go on so rapidly; and the temperature being the same, or very nearly so, when the water-bath is kept boiling, there is greater risk of injurious action from the air." — *United-States Dispensatory*, p. 930.

lastly, rubbing them in a mortar. The Edinburgh College direct that the seeds should be at first well softened with steam ; then sliced, dried, and ground in a coffee-mill."

PEREIRA prescribes the commencing dose at two or three grains, and pronounces its operation " very uniform." The " United-States Dispensatory " orders five grains, but declares its effects to be "*very uncertain ;*" inasmuch as fifty grains have been given in a single dose, with little or no effect.

What are we to think of such conflicting statements in our medical authorities? Simply this : That the few grains directed in the commencing dose are rightly stated, and that this is the dose which generally exhibits its " uniform " effects, — perhaps with slight variations in different constitutions, — *provided when thus administered it possesses its full activity;* and that the tenfold quantity, which produced " little or no effect," was destroyed by *steaming* and *store-drying.* If the attending physician, in view of the failure of the fifty grains to produce any effect, should give another dose of half that amount, dispensed by another apothecary,* who prepared his powder from the fresh nuts by " rasping " or " grating," *and not by first steaming to a pulp, and then by stove-drying or torrefying*, he would, in all probability, have no further occasion to prescribe for his patient. Under many of the present prescribed forms for preparing medicines for immediate use, how " very uncertain " are the instrumentalities frequently employed by the physician !

* See footnote, p. 7.

3*

It is difficult to speak of those syrups which are regarded as strictly medicinal, and which are formed on the basis of watery extracts, made by heat, except in close connection with the pharmaceutical operations already noticed. They are virtually decoctions, preserved in sugar : at any rate, the preservation of the decoction appears to be the great end in view; although an improvement in the taste of the compound is supposed to be secured by the addition of the sugar and the formation of the syrup.

But syrups are very liable to ferment and decompose. The quantity of sugar which should be added to a given measure of liquid has never been determined and given as a rule in the case. The books say, "About two pounds of sugar are usually required for one pint of liquid;" and perhaps this direction is as explicit as can be given. Sugar varies in its saccharine quality. Some kinds are stronger in their sweetening properties than others; and the variation in this respect adds to the difficulty of determining the exact quantity to be used. If the sugar in the syrup be deficient, fermentation ensues. If it be in excess, it is precipitated in the crystalline form; and these crystals, thus deposited, exert an affinity upon the saccharine matter still in suspension, gradually attracting portions of it, until, at last, fermentation ensues as surely as in the case where too small a quantity of sugar entered into the composition of the syrup.

Still, in face of all these difficulties of compounding an exactly balanced syrup, and the certainty which ensues of its speedy decomposition in default of such exact composition; and, further, notwithstanding the patent proofs, that, in the preparation of the basis of syrups, generally — namely, in the decoction — all, or very nearly all, of the virtues of the ingredients employed are dissipated and lost, — syrups have been the most popular of medicinal prescriptions for at least twenty-five years past. In this statement, we allude more particularly to those syrups which claim to include the full virtues of sarsaparilla. Figures would be almost tasked to express the number of bottles of "syrup of sarsaparilla" which have been sold since the days of SWAIM; and those who believe that these compounds, almost, if not quite, without exception, have been utterly inert, cannot but be astonished at the extent of credulity through which such an amount of worthless mixture has been sold.*

---

* "The *compound syrups*, if kept any time, are liable to various alterations, depending on the nature and degree of care used in their preparation. . . . When the sugar bears too small a proportion to the liquid, syrups are apt to run into fermentation. Even when the sugar is in proper proportion, the change often takes place if the solution contain much *amylaceous* or extractive vegetable matter. Even when too much concentrated, they may also undergo this change, from a part of the sugar being deposited in a crystalline state; and the crystals, attracting the sugar necessary to the preservation of the syrup, reduces its strength, and renders it liable to the same change as though it was originally too weak. . . . Various plans have been devised to preserve syrups; but the best is, to prepare them only in small quantities." — *Thomas's Universal Formulary*, pp. 519, 520.

"The medicated syrups are liable to undergo various alterations. . . . Syrups which contain too little sugar are apt to pass into the vinous fermentation, in consequence of the presence of matters which act as a ferment. Those which contain too much sugar deposit the same in the crystalline state. . . . The want of a due proportion of saccharine matter frequently also gives rise to mouldiness, when the air has access to the syrup. . . . It is obvious, that syrups which depend upon a volatile ingredient, or one readily changeable by heat [when once injured], cannot be restored to their

We shall confine our further remarks upon this subject to an examination of the nature and quality of some
of the prescribed forms of syrup of sarsaparilla ; and
we herewith subjoin a recipe for a simple form of the
same, taken from a current authority : —

" *Syrup of Sarsaparilla.*

"Take of sarsaparilla, fifteen ounces.
Boiling water, one gallon.
Sugar, fifteen ounces.

" Macerate the sarsaparilla in the water for twenty-four hours;
then boil down to four pints, and strain the liquor when hot ; after-
wards add the sugar, and evaporate to a proper consistence." —
London Dispensatory.

The Edinburgh process is the same as the above.
The Dublin College obtains, in the same manner, " four
pints of a concentrated, strained decoction ; and prepares
a syrup from this, according to the general direction of
the college." — *U. S. Dispensatory*, p. 1152.

In the above *formula,* we have the prescription for
the preparation of the *simple syrup of sarsaparilla* from

---

original condition. *At best, syrups are too apt to change;* and various mea-
sures have been proposed for their preservation. . . . The best plan is to
make them in small quantities. . . . ' Let the syrups be preserved in a place
where the heat does not exceed fifty-five degrees.' — *London Dispensatory.*
It would be difficult to comply with such a rule in this country." — *U. S.
Dispensatory*, p. 1144.

" As multitudes of distilled waters have been compounded from materials unfit to give any virtue, so numbers of *syrups* have been prepared from
ingredients, which, in this form, cannot be taken in sufficient doses to exert
their virtues; *for two-thirds of a syrup consist of sugar,* and the greatest part
of the remaining third is an aqueous fluid." — *New [London] Dispensatory,*
p. 359.

" Syrups are seldom active medicines, but are principally designed to
render others pleasant." — *American New Dispensatory,* p. 505.

four dispensatories. Here is the direction to "boil down"
from one gallon to four pints; and, after adding the
sugar, it is to be *further boiled* or *evaporated* to a proper
consistence. This boiling will do all that can be done
by this process, either to extract and retain, or to dissolve
and dissipate, the virtues of the root. We feel authorized to say, from the evidence which may be adduced,
and which we have in part already cited, that the entire
value of the root will be thus wholly dissipated in
vapor. Nothing will be left in the "four pints," save
the inoperative starch, mucilage, albumen, salts, and
possibly a slight portion of extractive tannin ; the last
nearly destroyed in quality, if not entirely decomposed.

But we desire to present evidence in this case from
the mouths of other and more competent witnesses.
By these, we think, we shall prove that a compound
prepared as directed above would prove wholly inert
and worthless. Considerable testimony, having an equal
application to the present case, was adduced in our
remarks upon the action of decoction upon vegetable
medicines ; and we herewith make further quotations
in the footnotes, having reference to the formula we are
now considering.*

---

* " This syrup is necessarily *a weak, if not inert, preparation;* the virtues
of the *sarsaparilla* being only partially extracted by water, at least by the
quantity of the *menstruum* ordinarily employed, *and being injured or destroyed by boiling.*" — *U. S. Dispensatory,* p. 1152.

" The root [sarsaparilla], if good, when chewed, produces a disagreeable,
acrid taste in the mouth. . . . The root is efficient in proportion as it possesses
this acrimony. . . . The virtues, *which are impaired by long boiling,* are ex-
tracted by alcohol." — *U. S. Dispensatory,* p. 636.

" The same result as to the superior efficacy of alcohol, as above, as a

Notwithstanding these admissions of medical authors, in some cases seemingly constrained, that the syrup of sarsaparilla, made on the basis of decoction, is inoperative, and that diluted alcohol does extract and preserve its virtues, *we find no recipe for a tincture of sarsaparilla in any dispensatory we have examined.*

In the different editions of the "United-States Pharmacopœia" are several prescriptions for a compound syrup of sarsaparilla, — some of them based on a decoction of the root and of the other component parts; and others which seem to partake, especially in the first directions for their preparation, of the nature of a tincture. The following is from the early editions of the "Pharmacopœia:" —

---

solvent for the acrid principle of sarsaparilla, has been obtained by the *French* experimentalists." — *Journal de Pharmacie [Soubeiran]*, xvi. 38.

"The activity of the root [sarsaparilla] is believed to depend on one or more acrid principles, soluble, to a certain extent, in water, cold or hot, and either *volatilized or rendered inert by chemical change at the temperature of two hundred and twelve degrees [boiling water]*. This fact appears to be demonstrated by the experiments of POPE,[†] HANCOCK,[‡] SOUBEIRAN,[§] BERAL, and others. . . . From sarsaparilla prepared in this way, I *found no sensible results upon any patient.*" — *U. S. Dispensatory*, pp. 906, 907.

"*Sarsaparillin* is probably the principle on which sarsaparilla depends chiefly, if not exclusively, for its medicinal powers." — *American Journal of Pharmacy*, vol. iii. 553–679.

M. BERAL states that he has formed this *sarsaparillin* pure by distilling a tincture of sarsaparilla made with very dilute alcohol. "In that case, it must be considered volatile; *and we can readily understand why sarsaparilla suffers by decoction.*" — *Am. Journal of Pharmacy*, xii. 245.

[†] Transactions of the Medico-Chirurgic Soc., Lond., xi 344.

[‡] Transactions of the Medico-Botanic Soc., Lond. — See also Phila. Journal of Pharmacy, vol. i. 295: "The observations of Dr. HANCOCK are entitled to much credit, as he practised long in South America, in the neighborhood of the best sarsaparilla regions."

[§] Journal de Pharmacie, tom. xvi. 38.

"Compound Syrup of Sarsaparilla.

" Take of sarsaparilla, ground in coarse powder, two pounds.
Guaiacum wood, rasped, three ounces.
Hundred leaves roses, ⎫
Senna, ⎬ of each, two ounces.
Liquorice root, ⎭
Oil of sassafras, ⎫
Oil of anise, ⎬ each, five minims.
Oil of partridge-berry, three minims.
Water, a sufficient quantity.
Sugar, eight pounds.

" Mix the five first ingredients with three pints of water, and
allow the mixture to stand for twenty-four hours. Then transfer
the whole to an apparatus for displacement,* and pour on water
gradually, until one gallon of filtered liquor is obtained. Evaporate this to four pints. Then add the sugar, and proceed to form the
syrup in the manner directed for syrups. Lastly, having rubbed
the oils with a small portion of syrup, mix them thoroughly with the
remainder." — *U. S. Ph.*

The above recipe varies considerably from that for
the simple syrup, previously copied. It is commenced
by a day's maceration in water ; then a quantity of water
is added, until, by " displacement," a gallon of filtered
liquor is obtained. Thus far, there is a seeming difference in the two preparations ; but the difference is
merely technical. Decoction or evaporation comes in at
last, and the liquor is vaporized down to four pints.
The slight extract of the virtues of the several ingredients obtained by the maceration, together with those
which remain in the substances employed before the
evaporation, are all dissipated by the boiling which fol-

---

* The process of *percolation*, or *leaching*.

lows; and a syrup thus compounded is, like the one previously spoken of, wholly inert. All the evidence before adduced to prove the worthlessness of the previous syrup, applies, in its full force, to this last *formula*.

So strong has been the conviction of late, that vegetable syrups, prepared by decoction of the compound medicinal substances in water, are altogether inoperative, that we find in some authors a prescription for them which purports to partake of the nature of a tincture. The primary boiling is omitted. The one we have copied above has given place to the following form, taken from the same work: —

"*Compound Syrup of Sarsaparilla.*

℞ "Sarsaparilla, bruised, two pounds.

Hundred leaves roses, }
Senna, } of each, two ounces.
Liquorice root. }
Guaiacum wood, rasped, three ounces.
Oil of sassafras, } of each, five minims.
Oil of anise, }
Oil of partridge-berry, three minims.
Diluted alcohol, ten pints.
Sugar, eight pounds.

"Macerate the first five ingredients in the alcohol for fourteen days; express and filter. Evaporate the tincture on a water-bath to four pints, filter, add the sugar, and form the syrup; then, having rubbed the oils with a little of the syrup, mix well with the remainder."— *U. S. Pharm. and Thomas's Universal Formulary.*

If we stay the composition of the above recipe at the word "filter," we shall have a compound tincture of sarsaparilla; which, on the supposition that all the articles submitted to the action of the diluted alcohol were of good quality, would undoubtedly prove efficient; and which, if properly secured in bottles, would remain in full virtue for an indefinite period. But the love of heat, or the mistaken estimation of its powers of extraction and concentration, must needs come in at last; by the operation of which, *all that was good in the tincture is diffused in the air by the subsequent evaporation.* Concentration, indeed! The reverse of this, in the present case, only is true; and, for concentration, we should read, *dissipation!* The alcohol *must be killed!* *

But the syrup formula is not quite so popular just now as it was in the first half of the present century. The recipes for these preparations, even in increasing numbers, are still found in all our dispensatories; but the conviction which at times has forced itself upon the minds of medical men, that they were of no value, has apparently given rise to new compounds, new in name at least, which are now denominated " FLUID EXTRACTS." These new forms, in some cases, are undoubtedly improvements; but there seem to be no acknowledged general rules by which they are composed. The apothecary appears to have little to do in this matter at present. So far as we can learn, the manufacturing of these extracts is taken from their hands, and placed in those less, responsible and capable.

---

* " The syrup of sarsaparilla, in the ' U. S. Pharmacopœia,' is intended to represent the *French sirop de cuisinier.* It is prepared with proof-spirit,. which extracts the acrid principle of the root, without taking up the inert fecula; and *the tincture, being evaporated* TO GET RID OF THE ALCOHOL,.is. made into syrup." — *Pereira's Mat. Medica,* vol. ii. p. 279.

A few of these extracts in use at the present time exhibit some of the characteristic powers of the articles from which they are made. We find in the shops fluid extracts of *valerian, ipecac,* and *hyoscyamus,* which partake more of the nature of tinctures, and less of the effects of heat; whose peculiar operative powers are clearly recognized : but, generally speaking, they are inert preparations.

We copy below a recipe for one of these new forms from the " U. S. Pharmacopœia : " —

"*Fluid Extract of Sarsaparilla.*

℞ " Sarsaparilla, bruised, sixteen ounces.

Liquorice root,
Guaiacum wood, } of each, two ounces.
Sassafras,

Mezereon, sliced, two drachms.

Diluted alcohol, eight pints.

" Digest for fourteen days; strain, express, and filter. Evaporate on a water-bath to twelve fluid ounces, add twelve ounces of sugar, and remove from the fire when this is dissolved." Dose one fluid drachm three or four times a day. — *Thomas's Universal Formulary.* *

The reader is desired to observe how nearly this recipe corresponds to the second formula we have copied of the compound syrup of sarsaparilla. The active articles employed, and the method of compounding them, seem almost identical. In both cases, a good tincture is first made, nearly of the same materials ; *and, by the second process of evaporation, is utterly destroyed!* One hundred and twenty-eight ounces of proof-spirit tincture boiled down to twelve ounces ! How much of the

---

* The only difference between this and the officinal formula of the U. S. Ph. is the omission of the guaiacum wood in the latter.

diluted alcohol may there be found in the twelve remaining ounces? Certainly, not one drop! Even more
than three-fourths of the water which formed the dilu-
tion is converted into vapor. It should be remembered
in this connection, that proof-spirit is allowed to be the
only proper and perfect solvent and preserver of the active, essential properties of sarsaparilla; and that, for
this purpose, its use, it would seem, is ordered in the
above prescriptions : yet it appears to be carefully provided that this should be all dispelled in vapor, and
with it all the virtue which the root contains! "Dose,
one fluid drachm"! It would not be pleasant, and
probably not safe, for a person to drink at once twelve
ounces of sugar in solution; but abating the sugar, and
the nauseous decomposition of the tannin and gummy
solution, a person of good digestion, who could bear the
draught of twelve ounces of water, or the same quantity of weak "herb-drink," might, it is believed, take
the twelve remaining ounces of this fluid extract with
entire impunity.

In the same "UNIVERSAL FORMULARY," we have a
recipe for the "*fluid extract* of *Virginia snake-root.*" *
The tincture of four pints is first made of the root,
which is subsequently evaporated to twelve ounces. In
this case, nothing remains of the snake-root, after the
completion of the process, except *mucilage, gum, earthy*

---

* "The tincture of snake-root possesses all the tonic and cordial properties of the root. . . . The *decoction* and *extract would be improper forms;*
as the volatile oil, upon which the medicine partly depends, is dissipated by
boiling." — *U. S. Dispensatory,* pp. 660 and 1186.

"By decoction [of snake-root], *its powers are entirely destroyed.*" —*American New Dispensatory,* p. 151.

matter, and possibly a partially decomposed astringent substance, of no value, and especially repulsive to the taste.

Some other forms omit the application of artificial heat in the preparation, and expose the alcoholic tincture to spontaneous evaporation of seventy-five per cent of the quantity employed; after which, the process is completed in the usual manner; the destruction of the tincture being thus accomplished in one week, instead of one hour.

We copy one more recipe in full : —

> " *Fluid Extract of Dandelion and Senna.*
> ℞ "Senna, two pounds.
>   Torrefied dandelion-root, one pound.
>   German chamomile, one-fourth pound.
>   Sugar, twenty ounces.
>   Oil of wintergreen, one-half drachm.
>   Carbonate of potass, or soda, one ounce.
>   Alcohol, two ounces.
>   Water, half a gallon.

" Powder the plants, and mix them with the water, holding the alkaline carbonate in solution. Let the mixture stand for twelve hours; then introduce into a percolator, and add water until a gallon of liquid shall have passed. Evaporate on a water-bath to twenty ounces; add the sugar; filter; and, when cold, add the alcohol, holding the oil of wintergreen in solution. Dose, a teaspoonful to a tablespoonful." — *Thomas's Universal Formulary*, " E. Dupuy."

The first article in this compound is senna, whose operative virtue is nearly or quite destroyed by the heat of boiling water, if continued even for a short time.* The second (namely, dandelion-root), when fresh,

---

* " It [the virtue of senna] is very conveniently extracted by water; *but it does not bear a boiling heat*, having much of its purgative quality thereby injured." — *Cullen's* [*Barton's*] *Mat. Medica*, vol. ii. p. 374.

abounds in a "milky juice," in which the medicinal quality of the root resides; which root must be carefully dried in the shade, to preserve its strength even for a few weeks.*

The third ingredient is *German chamomile*,† which, when fresh, has a strong and grateful aroma, so covering its bitterness as to render it an agreeable tonic, but which is deprived of its valuable medicinal properties by exposure for a short time to the air of a high summer temperature.

It is a conceded point, that the fresh dandelion-root should be used as soon after it is taken from the earth as is practicable; and that it cannot be preserved, even with the greatest care in drying, to such an extent as to render it proper for use beyond the following winter. The virtue of its milky juice is lost in a little time, despite of all care to preserve it. The *poppy-head*, which in its proper season exudes its milky juice through slight incisions; which, in the open air, is soon inspissated into opium, — if left to stand in the field until the flowers open and mature, loses all its narcotic juice, and the *capsules* thus dried are nearly as free of opium as the corn-husk. How much virtue, then, can we suppose would remain in the dandelion-root, which,

---

* Respecting the cure and the value of dandelion-root, the " United-States Dispensatory " says in substance, " If gathered before the close of warm weather, and dried in the shade, *with great care, its activity may be so far preserved, that it may with propriety be prescribed the ensuing winter.*"

† . . . " The aroma [of chamomile] is *dissipated by boiling.* The *decoction* is better calculated for fomentations and *enemata* than for internal use." — *U. S. Dispensatory,* p. 901.

as a preparatory step to its boiling, had been thus
" *torrefied* " ?

But, in the case we are considering, the result would
be the same if the roasting of the dandelion should be
omitted. The gallon of liquid obtained by the first
part of the process must be evaporated to one pint and
one-fourth. No professional acumen could detect the
difference, whether the torrefied or untorrefied root
were employed, or whether *any dandelion or senna*, and
perhaps *chamomile*, entered into the composition. The
process of compounding this fluid extract would nearly
or quite destroy all the prominent, valuable properties
of any or of all the most highly esteemed roots, leaves,
flowers, seeds, fruits, and rich juices, whose virtues
have given them a prominent place in our VEGETABLE
MATERIA MEDICA.

Such persistent efforts to dilute and destroy the remedial agents placed in our hands as a boon and
blessing ! Such seeming determination to reduce the
health-giving restoratives of the vegetable kingdom to
the weak wash of sweetened water, or to the inert
though nauseous extract of *amylaceous, albuminous,* and
earthy matter, each dose of which adds to the discomfort
and disgust of the patient ! But in this way they accomplish what seems to many medical men to be an object of
paramount importance ; for, in the words of a late author
already quoted, they thus " *get rid of the alcohol.*"

We have also inspissated, or, as they are sometimes
called, solid extracts, which, in the language of our
PHARMACOPŒIAS, embrace all products obtained by
the evaporation of a solution, maceration, digestion,

infusion, or of an expressed juice. They are of a dark color, of nearly the consistence of the *pill mass*, but gradually grow harder, especially on exposure to the air ; and some of them, in a little time, become pulverizable. They are generally made, when artificially produced, by prolonged decoction. The several vegetable substances are boiled until the *menstruum* becomes thick and dark : the liquor is then strained or filtered from the dregs, and subsequently evaporated to the desired consistence. Sometimes these extracts are made, as has been said, by commencing the process with maceration, digestion, or with the expressed juice ; inspissation being effected in a similar manner by continued boiling or evaporation.

Whichever of the above methods be adopted, the result is nearly the same. The valuable properties of the plants, subjected to the long boiling heat, are dissipated in the air. The extracts made in this manner all look very nearly alike, have a similar odor, and even taste ; and, in all probability, *a similar effect* when administered. The objections to extracts obtained by the process just described are recognized by many authorities ; and, to obviate them, the plan has been tried of obtaining them by boiling carried on *in vacuo*.* To this end, an apparatus has been formed by Mr. Barry.†

---

* "The objection to decoction and evaporation [in the formation of extracts] is obviated in a great measure by allowing the expressed juice to evaporate spontaneously in ordinary temperatures, or by carrying on the process *in vacuo*." — *Thomas's U. S. Formulary*, p. 517.

† "The apparatus for evaporating *in vacuo*, invented by Mr. BARRY, and described in the ' London Journal of Science and Arts,' enables the water to rise in vapor more rapidly, and at a comparative low temperature. The method of BARRY consists in distilling into a large receiver, from which the

The articles with their *menstruum* are placed in this apparatus modified by Mr. REDWOOD, in a close, strong vessel, and the air-pump applied to withdraw the atmospheric air and the vapor as it rises. It is evident there is no *vacuum* formed by this mode of operation, the steam rising as fast as it can be withdrawn by the pump; and it seems that only a *quasi vacuum* is claimed. But another claim is made ; namely, that the boiling goes on much more rapidly, and that the process is thus completed in less time, than is required without this mechanical withdrawal of the air and vapor. This is undoubtedly true. More of the volatilized *menstruum*, and of the essential qualities of whatever substances may have been put into the liquid, would be withdrawn, and diffused in the atmosphere, in a given time, than by the ordinary mode of ebullition. But it does not seem to be important to ascertain the difference of time required in the two cases. The effect must be the same in either case ; namely, the loss of the valuable principles of the medicines employed, in the " elastic incoercible vapor " which escapes. Thus the same objections

---

air has been expelled by steam, and in which the vapor is condensed by cold water, applied to the surface of the receiver, so as to maintain a partial *vacuum*. Mr. REDWOOD has modified this process by keeping an air-pump active during the evaporation; thus removing not only the air, but the vapor as fast as it rises."— *Journal de Pharmacie*, 3d series, 231.

NOTE.— The invention of Mr. BARRY, without the modification above described, is virtually embraced by the apparatus usually employed for the purpose of distillation. A liquid partaking of the virtues of the medicinal substances employed would thus be obtained, but no inspissated extract. The repetition, in this place, of some of the objections before made to this process *in vacuo*, seemed necessary, inasmuch as an improvement is here claimed, especially applicable to the formation of *solid extracts*.

apply to the preparation of extracts in this manner which have been already noticed in speaking of decoctions made by the same process.*

The methods generally prescribed by our *pharmacopœias* for obtaining the inspissated vegetable extracts are so various, and sometimes conflicting in their character, that they must greatly embarrass the *pharmaceutist.* The quotations in the footnotes, immediately following this reference,† are all from the " United-States

---

* In either preparation [of vegetable extracts, made by the boiling or the evaporation of a watery or spirituous solution], the volatile principles must necessarily be dissipated; and in many cases, especially in the preparation of watery extracts, decomposition, or oxygenation, of the more fixed parts takes place. Hence there are few vegetables whose virtues are obtained uninjured in their extracts.— *American New Dispensatory,* p. 538.

" Extracts are prepared by boiling the subject in water, and evaporating the strong decoction to a thick consistence. . . . There is a great difference in vegetable substances in regard to their fitness for this operation; some yielding all their virtues, and others scarce any. . . . *Those which contain a peculiar odor, flavor, and aromatic quality, are either not extracted at all, or exhaled along with the menstruum. . . . Wormwood,* which has a degree of warmth and strong flavor joined to the bitter, loses the two first in the evaporation. . . . An extract made from the flowers of *rosemary,* or *lavender, discovers nothing, either of the taste, smell, or virtues,* of *the flowers."* — *New* [*London*] *Dispensatory,* p. 403.

" In selecting a suitable *menstruum* in which to procure the principle of plants, reference is had to the nature of these. If they are gums or starch, which may be taken up by water, the cheap fluid is employed; and the *resins,* which are only soluble in alcohol or ether, are left behind. But, if it be the resins or the volatile oils which it is desirable to obtain, alcohol or ether is employed. . . . Various methods are adopted to effect solutions of vegetable principles, so as to obtain their full strength without endangering their decomposition by exposure to too great heat. It was the opinion of ORFILA, from numerous experiments upon extracts, *that their virtues diminish in proportion to the degree of heat to which they are exposed."* — *New American Cyclopædia.*

† . . . " *Long-continued evaporation has a tendency to injure the extract.*

" According to BERZELIUS, the injurious influence of atmospheric air is much greater at the boiling-point of water than at a less heat, even

Dispensatory," chapter " *Extracta ;*" some of which, however, are credited to other authorities. It will be seen that they are quite antithetical in several of their recommendations ; and it would seem to be a fair inference, that no one of them secures the general favor and confidence of medical men.

A mode of preparing extracts without artificial heat is briefly alluded to below ; but the process of preparation is prefaced with the doubt of a "*perhaps*," which gives little value to the paragraph of two lines.   The

---

allowing for the longer exposure in the latter case.   Therefore a *slower evaporation, at a moderate heat, is preferable to the more rapid effects of ebullition.* . . .

"The method of evaporation usually resorted to in the case of the aqueous solution *is rapid boiling over a fire.   The more quickly the process is conducted, the better,* provided the liquor is to be brought to the boiling-point; for the temperature cannot exceed this, and the length of the exposure is diminished. . . .

"It has been proposed to effect the evaporation at the common temperature by *directing a strong current of air, by means of a pair of smith's bellows,* over the surface of the liquid. . . .

"Plans have been proposed, and carried into execution, for performing evaporation *without the admission of atmospheric air.* . . .

"In the preparation of extracts, evaporate the moisture *as quickly as possible,* in a broad, shallow dish, by means of a water-bath. . . .

"Extracts are usually prepared by evaporating the expressed juice of plants, or their infusions, or decoctions in water, or proof-spirit, *at a temperature not exceeding* 212°, by *means of a vapor-bath.* . . .

"And the extracts of expressed juices cannot, perhaps, be better prepared than by spontaneous evaporation in shallow vessels, exposed to a current of air. . . .

"All simple extracts, unless otherwise ordered, ought to be prepared according to the following rule: Boil the vegetable matter in eight times its weight in water, *till the liquid is reduced one-half.*   Then express; and, after the subsidence of the dregs, filter.   Evaporate the liquor with a superior heat (between 200° and 212°), until it begins to thicken.   Finally, inspissate it with a medium heat (between 100° and 200°) obtained by a water-bath, frequently stirring until it acquires a consistency proper for the formation of pills."

final and explicit order, it will be seen, is, — in the commencement of the preparation, — Boil down the menstruum "until the liquid is reduced one-half."

The difference between extracts formed by decoction and evaporation, and those obtained from expression, or exudation, and inspissated, without the aid of artificial heat, is strikingly illustrated by the process of obtaining the narcotic property of the *poppy*, in contradistinction to the methods usually employed to obtain the extracts of medicinal plants.* If the head of the poppy, at the proper season of the year, be slightly incised, a milky juice immediately exudes, which soon hardens in the air into nearly perfect *opium*. Perhaps one grain of the extract may be thus obtained from each plant. If a quantity of poppy-heads be taken from the same enclosure, and subjected to the usual process of obtaining extracts by boiling and evaporation, a large quantity of extractive matter is obtained, amounting, it is said, to more than one-third of the weight of the heads used. This extract is composed of fixed oil, starch, albumen, earthy matter, &c., but contains no appreciable quantity of opium.†

---

* "The method of collecting the juice of the poppy is the following: At about three or four of the clock in the afternoon, individuals repair to the fields, and scarify the *capsules* with sharp iron instruments. . . . The capsules having been scarified in the manner above described, the collection of the juice is made at an early hour in the following morning. . . . The opium now requires frequent attendance on the part of the cultivator. *It is daily exposed to the air, but* NEVER TO THE SUN; *and is regularly turned over every few days, to insure uniform dryage in the whole mass.*" — *Pereira's Mat. Med.*, art. "WHITE POPPY."

† M. BERAL observes, in relation to the extract of poppies, as follows: "If made over an open fire, it is nearly inert."

The solid extracts should be made without the application of artificial heat, from the expressed juice,
exposed in proper vessels to spontaneous evaporation.
The peculiar odor of the several plants is thus preserved; and, what is far better, the physician witnesses
the effects of the peculiar properties ascribed to each
plant from their administration. In this manner, substantially, does Nature form nearly all our most valuable
gum-resins; namely, from the spontaneous exudation
of their juices, which are hardened in the temperature
of the atmosphere.

Comparatively few vegetable medicines are given in
substance or in *powder*.* Some are nauseous and repulsive in this form; some are too bulky to be taken in a
proper quantity in powder; some are nearly destroyed;
and all are more or less injured by drying them

---

* "This form [that of powder] is proper for such materials only as are
capable of being sufficiently dried for pulverizing without the loss of their
virtue. There are many substances, however, of this kind, which cannot
be conveniently taken in powder. Bitter, acrid, and fetid drugs are too
disagreeable; emollient and mucilaginous herbs and roots, too bulky.
Pure gums cohere, and become tenacious in the mouth. Fixed alkaline
salts deliquesce when exposed to the air, and volatile salts exhale. Many
of the aromatics, too, suffer great loss of their odorous principle when kept
in powder, as in that form they expose a much larger surface to the air." —
American New Dispensatory, p. 541.

"If powders are long kept, and not carefully secured from the air,
their virtue is in a great measure destroyed, although the parts in which it
consists should not, in other circumstances, prove volatile. Thus, though
the virtues of *ipecacuanha* are so fixed as to remain entire even in extracts
made with proper *menstrua*, yet, as the College of Wirtemberg observes, if
the powdered root be exposed for a length of time to the air, it loses its
emetic quality." — *New [London] Dispensatory*, p. 548.

"Powders of aromatics are to be prepared only in small quantities at a
time, and kept in glass vessels very closely stopped." — *Pharmacopœia
Edinburgensis.*

sufficiently for pulverization. Most prescriptions, consisting of powders, are designed for acute or febrile cases, and are seldom used in chronic forms of disease. *Opium, ipecac,* and *camphor,* in powder variously compounded or mixed, form a very common *febrifuge; rhubarb* and *jalap* are often, though less frequently, given in powder; and some other articles might be named, which are occasionally ordered in this form. But it is believed that all vegetable medicines lose a considerable portion of their active properties by drying to the degree required for pulverization. Their loss in weight by this process is well known. Thus *rhubarb* loses in weight about eight per cent; *columbo,* ten; *valerian,* fifteen; *Virginia snakeroot,* twenty; *ipecac,* twenty-five; *squills,* eighteen; *cinchona,* twelve; *cinnamon,* twelve; *angustura,* eighteen; *chamomile flowers,* fifteen; *saffron,* twenty; *lavender* and *rosemary,* eighteen; *digitalis* and *belladonna,* twenty; *senna,* twenty-five; *henbane,* forty-five; *colocynth,* fifty: (see tables of GUIBOURT and HENRY.) The loss in the purgative and aromatic gums is comparatively light. A part of this deficiency, especially in the case of some of the roots, is undoubtedly owing to the presence of a ligneous matter, which is of little value, and scarcely pulverizable; but another portion is obviously occasioned by the preparatory drying process, in which, whatever care may be used, much of the valuable properties of the several articles escapes: so that the loss in weight, as indicated above, may pretty nearly represent also the loss in virtue.

Conserves, troches, confections, and *electuaries* have been little used as medicinal forms for many years; and,

5

at the present time, seem to have fallen into entire·
disuse.

Certain remedies are best administered in *pills;* for,
by this mode of preparation, the offensive taste of
the component parts is almost wholly concealed. It
is not, however, so well adapted to the demands of
acute diseases where an immediate effect is desired ;
yet its slower solution and operation are not objected
to in the treatment of chronic complaints. Some
articles are too bulky for this mode of preparation ; but
there are quite a number of active materials in very
common use, sufficiently concentrated for this purpose.
Among these are the purgative gum-resins and narcotic
extracts. The *pill mass* is also a convenient and proper
form for *expectorants,* as of *squills, ipecac, antimony,* &c. ;
and most necessary for the administration of the *fetid*
gums, such as *assafœtida, galbanum,* and *sagapenum.* *

All, or nearly all, of the simple medicines of the
vegetable kingdom, yield their virtues to alcohol or
proof-spirit ; and they not only yield these to this solvent, but, with very little care, their efficiency is *fully
and permanently preserved.* " Proof-spirit, bottled, and
corked moderately tight, *never ferments.*" This declara-
tion is from a source entitled to full credit. The rule
is, that *tinctures* never lose any portion of their virtue by
age. There may be exceptions to this rule, which is
here more particularly intended to apply to the medici-

---

* " No form of medicinal agents is more frequently employed than that
of *pill;* not only because of the facility with which it is administered, and
its comparatively little taste, but because this form answers so excellent a
purpose in the preservation of certain compounds." — *Universal Formulary,*
p. 515.

nal agencies of the vegetable kingdom; but they are very few. In the language of our *dispensatories*, they preserve their virtues "*for an indefinite time.*"

The use of tinctures was formerly more extensive and popular than it is at present. It is admitted, now as formerly, that they present the active principles of drugs in a small volume; and that they are, for other reasons, proper forms of prescription.* Some two or three hundred *recipes* for tinctures are contained in our late dispensatories: still, while these authors commend this form of preparation as one which fully extracts and permanently retains the virtues of those agents submitted to the influence of spirituous solvents, they clearly interdict their general use,† for reasons we shall presently consider.

If it can be clearly shown that spirit-tinctures dissolve more perfectly the valuable properties of vegetable medicines than any other prescribed mode of extraction, and that they preserve these in full activity and efficiency for an unlimited period, and are the form of exhibition most agreeable to the patient, especially in protracted diseases, the evidence should re-

---

* Rectified spirit of wine is the direct menstruum of the resins and essential oils of vegetable matter; and totally extracts these active principles from sundry vegetable matters, which yield them to water, *either not at all, or only in part.* — *New Dispensatory,* p. 303.

"The form of tincture is one much used in *pharmacy.* It presents the active principles of *drugs in a small volume. It can be preserved in an unaltered state for a long time;* and it is, in most cases, well adapted to unite with other substances in extemporaneous prescriptions." — *Universal Formulary,* p. 521.

† "Physicians should avoid prescribing alcoholic remedies in chronic cases, whether alone or in the form of tincture, for fear of begetting intemperate habits in their patients." — *U. S. Dispensatory,* p. 63.

instate them in general favor. They should take the
place of those forms of prescription which sometimes
decompose the plants employed, which at best dissolve
less of their active properties, dissipating even these
in vapor, or retaining their greatly diminished virtues
but for a short time.

We will present some further testimony from approved writers in favor of the use of tinctures, as
affording a most eligible form for the preparation and
exhibition of vegetable remedies.*

---

* " The powdered root [valerian] impregnates both water and alcohol.
Water distilled from it smells strong of the root; but no essential oil sepa-
rates, whatever be the quantity employed. The watery extract is strong
and disagreeable, sweetish and bitter; but the spirituous agreeable, and
nearly resembling the root . . . Next to the powder, a strong tincture
made with proof-spirit is the most efficient." — *Parr's Medical Dictionary.*

" *Canella yields all its virtue to alcohol.* Boiling-water extracts nearly
one-fourth of its weight; but the infusion, though bitter, has comparatively little of the warmth and pungency of *canella.*" — *U. S. Dispensatory,*
p. 159.

" As the virtue of *valerian* resides chiefly in the volatile oil, the medicine
should not be given in decoction or extract. . . . The tincture is officinal."
— *U. S. Dispensatory,* p. 732.

" The tincture possesses all the tonic and cordial properties of the root
[snake-root]." — *United-States Dispensatory,* p. 1186. . . . " The *decoction or
extract would be an improper form.*" — *U. S. Dispensatory,* p. 660.

" Water extracts the virtue of sundry fragrant, aromatic herbs: . . .
but the aqueous infusions are far from being equally suited to this process
with those made in spirit; *water carrying off the whole odor and flavor, which
these higher liquors leave entire behind it.* Thus a watery infusion of *mint loses
an evaporation the smell, taste, and virtues of the herb; while a tincture drawn
with pure spirit yields a thick, balsamic liquid, or solid gummy resin, extremely
rich in the peculiar qualities of the mint.*" — *New* [*London*] *Dispensatory,* p. 49.

" The virtues of many vegetables are extracted almost equally by water
and rectified spirit: but, in the watery and spirituous extract of them, there
is this difference, — that the active parts in the watery extraction are blended
with a large portion of gummy matter, on which their solubility in this
menstruum depends; while rectified spirit extracts them almost pure from
gum." — *New* [*London*] *Dispensatory,* p. 303.

We think the references made in this paper to many approved authorities, in relation to "simple medicines," prove several facts worthy of the careful consideration of the physician. These are very briefly repeated here, as follows : First, that the gathering and curing of these articles is a matter attended with considerable difficulty ; second, that their preservation is imperfectly secured with the exercise of still greater care ; thirdly, that, despite the best efforts to collect in a proper manner and in the right season, and to preserve, plants thus gathered, some begin to decay at an early day, and, in many instances, soon thereafter become inert and worthless, yet still, perhaps, preserving a good outward appearance.

We quote substantially from our *dispensatories,* and systems of *Materia Medica,* in saying that a large proportion of the *sarsaparilla* * in our market is inert ; that *colchicum-root,* as usually imported, is worthless ; that our *snake-root, valerian, cubebs, buchu, chamo-*

"The active principles of dried vegetables can only be extracted by means of a liquid solvent. The *menstruum* usually employed is either water or alcohol, or a mixture of the two. . . . Alcohol is employed when the principles to be extracted are insoluble, or slightly soluble, in water, as in the case of resins; when it is desirable to avoid in the extract *inert substances, such as gums or starch ;* when the heat required to evaporate the aqueous solution would *dissipate or decompose the active ingredients of the plants,* as the volatile oils and the active principle of *sarsaparilla ; when the re-action of the water itself upon the vegetable principles is injurious ;* . . . and, finally, when the nature of the substance to be exhausted requires so long a maceration in water as *to endanger spontaneous decomposition.* The watery solution requires to be soon evaporated, *as the fluid rather promotes than counteracts chemical changes; while an alcoholic tincture may be preserved unaltered for an indefinite period.*" — *United-States Dispensatory,* pp. 926, 927.

* See a previous footnote, relating to the quality of the sarsaparilla of the shops, p. 12.

mile flowers, mezereon, pink-root, many medicinal *barks, leaves,* and *seeds,* are frequently much impaired in virtue before they are wrought into their ultimate forms of use by the apothecary or physician. But the most important fact, which we would again commend to consideration, is the frequent admission, as we have shown, of medical writers, that the forms of preparation before noticed, as made with the assistance of external heat, *even if the best materials are used, do not dissolve and retain the virtues of the ingredients employed.*

Now, in view of these frequently admitted facts, does not a very important question force itself on our consideration? Is there no method, or form of preparation, by which the most active medicinal, vegetable plants can be taken when fresh, and *known to be in their full virtue,* and so compounded as to *secure and retain their full efficiency for an indefinite time?* We think we have already shown, from as good authority as there is extant on this subject, that there is such a form ; namely, the form of tincture, made from alcoholic or proof spirit. We will offer here no further evidence on this particular point, but would ask a careful consideration of the proofs that have been already presented ; and which may be summarily expressed as follows : —

That diluted alcohol, or proof-spirit, is the direct solvent of all the active medicinal agencies of the vegetable kingdom ; while water, hot or cold, extracts them either partially, or not at all.

That the spirit-tincture not only dissolves the entire virtues of the most esteemed articles of the vegetable *Materia Medica,* but that it preserves their strength in full force, and in a permanent form; whereas the partial value of the solutions made by water is with great difficulty retained even for a short time.

That the form of tincture affords an eligible prescription for use. Its administration does not, like some of the preparations we have noticed, cause *nausea*, thereby increasing the discomfort of the patient; but almost invariably tends to allay all unpleasant sensations of that character which may exist at the time of its use.

That spirituous tinctures extract from medicinal vegetables their active properties only, leaving entirely untouched the inert principles of the plant, such as starch, albumen, gum, and earthy matters, which can scarcely be called medicinal, but which are taken up by the aqueous preparations we have already noticed, the use of which soon becomes offensive to the taste, and frequently produces a disturbance in the stomach, which greatly retards the recovery of the patient.

That tinctures present the active principles of drugs in a small volume; which is frequently a matter of great importance, inasmuch as some prescriptions, from their mere bulk, offend the stomach, and thereby produce a general derangement of the system.

It would seem that all these conceded advantages in favor of the use of tinctures would secure their general employment, especially in diseases of a chronic character, or in all cases in which alterative remedies are indicated; but the use of tinctures at the present time is regarded with nearly general disfavor. If any alcohol be introduced in the early process of the composition of a prescription to extract the active principles of the

several ingredients which are ordered, the closing directions frequently are to boil down the liquid obtained
from fifty to eighty per cent, in order, it would seem,
to "get rid of the alcohol;" although, by following
the last order, the pharmaceutist will effectually dissipate
all of the essential operative powers obtained in the
first instance by its employment.

In this connection, there seems to be a necessity for
some remarks on the medicinal character of alcohol, or
of proof-spirit, and of vinous liquors. Spirits, as we
have said, are not in favor, even as a vehicle or solvent
for the virtues of plants. Some take the ground, that,
inasmuch as their immoderate use has been produc-
tive of so much degradation and misery, they should
not be used in any case or under any circumstance.
Some prescribe their use *per se;* others acknowledge
their good effects, when they hold in solution appropriate medicinal remedies, and in this form recommend
their use. Thus opposite opinions * are entertained upon

---

* " Concerning the use of alcoholic liquors as a medicine, the ablest
writers hold directly opposite opinions, while they all of them agree in warn-
ing against the dangers that are sure to attend their habitual use. The best
authorities treat of them as a stimulant, that may be employed to advantage in sustaining the enfeebled powers in advanced stages of fever, or in
hastening the restoration in convalescence from acute diseases, and also
in cases of chronic debility. For *indigestion, colic, lock-jaw,* and some other
diseases of a violent and sudden character, alcohol may prove of great
benefit. . . . Yet many would be glad to see it banished from the *Pharmacopœia,* in view of the immense evils it brings upon the unfortunate
individuals who acquire a taste for it." . . . *New American Cyclo-
pœdia.*

" The advantages of *wine,* as a *pharmaceutical menstruum,* are, that, in
consequence of the alcohol it contains, it dissolves substances insoluble
in water, and, to a certain extent, resists the tendency to spontaneous
change. . . . But most wines, particularly the light varieties, are liable to

this subject at this day; although the number who repudiate the use of spirit, either unmixed or as a medicinal solvent, are probably in the majority.

---

undergo decomposition; . . . so that medicated wines, although they keep *much better than infusions or decoctions, are still inferior, in this respect, to tinctures.*" — *United-States Dispensatory,* p. 1209.

"The spirituous liquors, in small quantities, prove a powerful cordial, and, for a time, a strengthening beverage; giving vigor to the stomach, promoting digestion, and preventing flatulence. To the weakly and relaxed they are highly useful, by giving elasticity and firmness of tone." . . . *American New Dispensatory,* p. 135.

"As a pharmaceutic agent, it is of much importance, from the solvent power it exerts over a number of vegetable principles; . . . and by its property, too, of counteracting the spontaneous changes to which vegetable matter is liable." — *American New Dispensatory,* p. 72.

"In some states of acute disease characterized by excessive debility, it [alcohol] is a valuable remedy. In the form of brandy, it is frequently given in the sinking stages of *typhus* with advantage. Other kinds of ardent spirits are occasionally administered, and each is supposed to have its peculiar qualities. Thus, according to PARIS, *brandy* may be considered as simply cordial and stomachic; *rum,* heating and sudorific; *gin* and *whiskey,* diuretic." — *United-States Dispensatory,* p. 63.

"As a stomachic stimulant, spirit is employed to reduce spasmodic pains and flatulency, to check vomiting (especially sea-sickness), and to give temporary relief in some cases of indigestion attended with pain after taking food. As a stimulant and restorative, it is given with considerable advantage in the latter stages of fever. As a powerful excitant, it is used to support the vital powers, to prevent fainting during a tedious operation, to relieve *syncope* and languor, and to assist the restoration of patients from a state of suspended animation. In *delirium tremens,* it is not always advisable to leave off the employment of spirituous liquors at once, since the sudden withdrawal of the long-accustomed stimulus may be attended with fatal consequences. . . . In mild cases of diarrhœa, attended with griping pain, but unaccompanied by any inflammation, a small quantity of spirit and water, taken warm with nutmeg, is often a most efficacious remedy." — *Pereira's Mat. Med.,* vol. ii. p. 910.

"Wine is employed medicinally, chiefly as a cordial, stimulant, and tonic. . . . In the latter stages of fever, when languor and torpor have succeeded to a previous state of violent action, and in the low forms of this disease, wine is at times undoubtedly useful. It supports the vital powers, and often relieves *delirium, subsultus tendinum,* and promotes sleep. . . . As

For a large and highly esteemed class of vegetable remedies in general favor with the profession, no other menstruum so perfectly dissolves, and certainly pre-

---

a stimulating tonic and invigorating agent, it is given in a state of convalescence from fever, and in the recovery from various chronic non-febrile diseases." . . . — *Pereira's Mat. Med.*, vol. ii. p. 896.

"Spirituous liquors, in small quantities, prove a powerful cordial, and, for a time, a strengthening beverage; giving vigor to the stomach, promoting digestion, and preventing flatulence." — *American New Dispensatory*, p. 135.

"Advantage is sometimes experienced from stimulating drinks, as a little brandy and water during meals; but the moral hazard is, on the whole, greater than any probable benefit." — *Wood's Practice* (5th Phila. ed.), vol. i. p. 587.

. . . "On ne peut qu'approuver un précepte que GALIEN nous a conservé de ce médicin. C'est de donner du vin pour dissiper les roideurs qui se font sentir après les grands evacuations. C'étoit dans la même vue qu'HIPPOCRATE conseilloit de boire du vin pur, de tems en tems. . . . DISCORIDE et AVICENNE, après HIPPOCRATE, ont dit qu'il étoit utile, pour le santé, de boire, quelquefois, jusqu'à se enivrer: il est assez naturel de penser que, pour affirmir sa constitution, on pourroit se permettre, quoique rarement, des excès, autant dans le boire que dans le manger." — *Encyclopédie des Sciences*, par M. DIDEROT.

NOTE. — The permission "*rarely*," given above by the "*fathers of medicine*," would find few indorsers at the present day.

"Those [tinctures made with diluted alcohol] which require to be given in large doses should be cautiously employed in this form, lest the injury done by the *menstruum* should more than counterbalance their beneficial operation. This remark is particularly applicable to chronic cases of disease, in which the use of tinctures is apt to result in the establishment of fatal habits of intemperance." — *United-States Dispensatory*, p. 1160.

NOTE. — The amount of tincture which the above authority considers a "large dose," on account of the diluted alcohol it contains, may be seen in the following extract: —

"This tincture [of *valerian* made with diluted alcohol] possesses the properties of valerian, but cannot be given in ordinary cases, so as to produce the full effects of the root, without stimulating too highly, in consequence of the large proportion of spirit. The dose is from one to four fluid drachms." — *United-States Dispensatory*, p. 1187.

₊ The quantity above directed, expressed by a more common measure, would read, "Dose, from one-sixteenth to one-fourth of a wine-glass."

serves, their virtues as alcohol. There can be no need of saying more on this point; for it stands forth in our medical authorities as a conceded fact. Still, spirituous tinctures are frequently denounced, because their use may lead to intemperance.

Where an appetite for the use of ardent spirit has been formed, certain *quasi* medicinal liquors are frequently availed of, under the pretence that they are used " as a medicine," to gratify the *previously acquired love* for spirituous liquors. " *Spice bitters* " were formerly for sale in almost every shop in the country. Each paper generally contained one or two drachms of powdered cassia, with the make-weight, perhaps, of hemlock-bark, " sufficient for one quart of spirit." Other things of a kindred nature have wrought great mischief among us. " WOLFE'S SCHNAPPS," claiming to be a medicated spirit, have had a very extensive sale. This preparation is simply *gin*, generally of an ordinary quality, flavored, not entirely with oil of juniper; some two or three drops of other essential oils being added to each gallon.

But is there the man now living who dates his first love of spirituous liquors to the occasional and proper use of those formerly most common and popular compounds, — *hiera picra, elixir proprietatis, elixir sacrum, tincture of blood-root, tincture of guaiacum,* or *wine of ipecac?* I have never found one such person, — have not been able even to hear of such a case.

Tinctures fully saturated with active medicinal agencies cannot be taken in doses much over one ounce; and their prescribed quantity does not often exceed one-

half that quantity, frequently not over two fluid drachms.
If a larger amount than the maximum dose above stated
be taken for the sake of the alcoholic vehicle, it would
probably operate much like the "*remedy for intemperance,*" which had a transient popularity some thirty
years since; that is, it would overtask the stomach, thus
creating a disgust, instead of increasing the appetite, for
spirituous liquors.

The skilful, faithful physician has other important
duties to perform beyond those of prescribing appropriate
remedies in their best forms to his patient. If spirituous liquors, *per se*, have been ordered to meet any emer-
gency, which, in his opinion, required their use, he
should, as soon as the symptoms show any remission in
the severity of the case, diminish their quantity, and,
as soon as possible, discontinue their use entirely. He
should do this, not only to prevent the appetite for
stimulating drinks which might arise from their unne-
cessarily prolonged use, but that he may save for his
patient his susceptibility to the favorable operation of
the same remedy, should a recurrence of the disease
render their future prescription necessary; for a long
use of any remedial agency renders the system less and
less susceptible to its influence.

Opium, in the hands of the skilful and discreet *physician*, is among the highest gifts which our *Materia
Medica* confers on suffering humanity. How often,
through its proper administration, have the severest,
most excruciating pains been alleviated! The most
violent cramps and tetanic spasms are frequently relieved by the exhibitions of a few doses of opium; and

time has thus been gained for the employment of other remedies of slower operation ; by the use of which, permanent relief has often been gained. But the abuses of opium are widely known, although the baneful and degrading effects of its constant and excessive use are, we believe, more frequently exhibited in Europe and Asia than in the United States. Perhaps, in the perversion of this excellent remedy to all the evils flowing from intemperance, the physician may have sometimes incurred a fearful responsibility. If the patient be allowed to continue the use of opium beyond the absolute necessities of his case, a rising appetite for its stimulus may arise, which at length may defy the advice of physicians and friends, and even the sternest resolution which the subject of its influence may be able to form. The physician who prescribes an appropriate remedy at the proper time, and orders its disuse when it is no longer necessary, exhibits in each case, equally, his good judgment ; and, in each case, confers upon his patient the like benefit.

Quinine is a remedy highly and universally esteemed. In certain complaints, it is valued above any other, and perhaps all other, remedies ; yet its use, there can be no doubt, in too many instances, has degenerated to abuse. When taken in too large quantities, or continued beyond the necessities of the disease for which it was given, it has superinduced a derangement of the digestive organs, and an irritable nervous temperament of a most obstinate character.

Etherization, or the process by which a person is temporarily brought into a state insensible to pain by the inhalation of ether, that he may be better enabled to endure a surgical operation, seems, to the natural apprehension at least, to be the highest gift which could be vouchsafed to man under the experience of the severest pain. But the abuses of the use of ether are already known, even of the grossest character; and there is great reason to fear that such misuse of its influence is fast increasing.

We will not mention particularly other cases of the use, sometimes abused, of other remedial agents; but are ready to admit in general terms, that the perversion of use is incident to every medicinal agent worthy of a place in our dispensatories, and esteemed by the *Faculty*. The highest and most beneficent influences wrested to the worst purposes, or good degraded to evil, may be too often witnessed in every department of life, civil, moral, and religious. The great questions propounded for our solution, in all these cases, are, first, whether the abuse of good is, in any wise, a necessity of its use. All candid and fair-minded men must answer this question in the negative; although there are some, — who, it would seem, are neither impartial nor intelligent, — who, by their peculiar casuistry, discountenance certain high uses, on account of their possible, perhaps we should say frequent, perversion. Let us, in all cases, accept the use, and, as far as possible, eschew the abuse. Good of every kind was designed for a blessing. It flows freely and continually into every open form of use, into every instrumentality which can confer a benefit on man. A good thing can become evil, only by a perversion of its inmost beneficence to selfish and base purposes. The abuse of a good influence *is not a necessity*, but a wanton perversion of the gifts it ever seeks to bestow. The second interrogatory — namely, whether it is proper, *in any case*, to decline the good, from a fear that it may be wrested in our hands to evil — should also be promptly answered in the negative. It is our duty to receive every good gift which comes to us, co-operating in our freedom, to the end that it may accomplish in us the purpose whereunto it was sent; ever avoiding, as far as possible, all perversion of its influence.

But we do not propose to dwell on the subject of *ethics*, except so far as a great moral principle is involved in the matter under consideration; and we will now close our paper by offering a very brief *resumé* of the principal points or facts we have herein endeavored to establish, more by the authorities we have cited in the footnotes, than by our own convictions or arguments in the matter.

It is admitted that *infusions, decoctions, syrups, fluid* and *solid extracts*, compounded by the assistance of artificial heat, do not dissolve, and retain in a permanent form, *any* of the most active and valuable properties of the medicinal substances of the vegetable kingdom; even the astringency of plants, which is more readily extracted by water, being either entirely "decomposed," or, at best, "greatly injured by the boiling and evaporating process." This seems to be conceded as the rule which applies to these forms of preparation. To this rule, there are, without doubt, some exceptions: —

That the convenient and proper exhibition of remedies of the vegetable class, in substance or in powder,

is confined to few articles comparatively ; and that even
these few, under the greatest care, suffer a considerable
loss of their active qualities by the necessary process of
preparing them for powdering ; and, in some instances,
by the process through which they are reduced to this
form.

That the pill mass presents a better and more retentive
form for a certain (though quite limited) number of
remedies, expecially for the purgative gum-resins and
the active narcotic extracts.

That "the simple remedies" cannot be permanently
preserved in their full activity in an uncompounded
state, even under the greatest care ; and that, without
such care, they soon decline in value, and frequently
become worthless ; in many cases, still preserving the
external appearance of their original value.

That when the "simples" are taken even in full
strength, and compounded in an aqueous menstruum, by
the continued appliance of heat, as directed in the *for-
mularies*, their valuable properties are dissipated in
vapor : thus the result of the composition would be the
same, *whether active* or *inert ingredients were used.*

It is admitted that proof-spirit *will extract all of the
essential medicinal virtues of plants*, and will preserve
the same in full strength *for an indefinite period.* Age
improves tinctures, and ages will not injure them ; and,
if there be any exceptions to this rule, they are no
more than sufficient to establish its truth.

Thus it appears that the physician may always have
at command a class of remedies of whose operative
powers he is well assured. The apothecary also, if he

employ good and fresh materials in the composition of his tinctures, may dispense the same, with the confident assurance that they contain all the valuable and healing properties which the component parts possessed when gathered in their proper season, and found, on deliberate examination, to be in their highest state of perfection.

We have also endeavored to show, not from specific citations of proof in the case, but from collateral circumstances which seem to have a legitimate bearing upon the subject, that the use of tinctures, *fully saturated with active medicinal principles*, will not — cannot, it would seem — beget a rising appetite for spirituous liquors. Strong medications, even in those forms best accommodated to the taste, are ever more or less unpleasant, and increasingly so in proportion to the continuance of their use; which use almost invariably begets a disrelish or disgust, instead of a love, for spirituous liquors.

We would now submit, that the main facts and principles we have herein presented, which relate principally to the collection and preservation of vegetable medicines, and especially to the most eligible and efficient forms of their prescription, are well established by the evidence we have adduced from current, standard authorities. A part of this evidence, we are ready to allow, partakes somewhat of the character of reluctant admissions, but which, perhaps, are not the less valuable because they are unwillingly made.

If our premises have been sustained, what are the obvious duties of the physician? He should, at his leisure, carefully select his remedies, which are found,

on proper examination, to be fresh and in full force, and
immediately prepare the same for future use in such a
manner as will best dissolve and retain their virtues.
Thus he will ever have at hand remedies whose active
powers he well knows; and thus will he possess a primary advantage over those who frequently, in the hurry
of the moment, avail of an extemporaneous prescription,
compounded, perhaps, under the disadvantages of haste,
and possibly from materials which have declined in
virtue, yet still preserving a sound and good exterior
appearance.   Nor will the conscientious physician, who
appreciates the advantages frequently arising from the
preparation and prescription of alcoholic tinctures, be
deterred from their use by the persistent opposition,
open or covert, which has been made to their employment.   The doctrine which of late has been so generally
inculcated, that some of the good gifts which come
down to us, striving, as it were, to flow forth into the
ultimate forms of human blessing, must be declined, lest
the reception of their influence might be perverted to
evil, is a pernicious doctrine, of presumptuous import,
and entirely unworthy of a truly philosophic and Chris-
tian age.